by
Émilie BEAUMONT

Illustrated by
Lindsey SELLEY

· DISCOVERING ANIMALS ·

FOREST ANIMALS

by
Émilie Beaumont

Illustrated by
Lindsey Selley

TRODDY
BOOKS

The Red Deer

These animals live together in a herd or group. They are woodland animals, but also live happily in parks. The older males, called stags, often live on their own. Stags have big antlers, or horns, with up to twelve points. They grow new antlers every year. The females, called hinds, do not grow antlers.

The young calves are born in the spring. They cannot walk very well at first. The mother sometimes leaves her calf well hidden while she goes off to feed on grass and leaves.

The roar of the stag

In the autumn, the stag lifts his head and makes a loud roar. You can hear him a mile away. He roars to tell the hinds what a fine stag he is, and to warn other stags to keep away.

A duel

In the autumn, the stags fight over the hinds. These fights are not serious, because the weaker stag runs away when he feels he cannot win.

The young deer

The coat of a young calf is spotted. This makes it harder to see when it is lying hidden in the woods. Later it grows a new coat like its parents. As soon as the young deer learns to walk and run, it starts to play with other calves. It takes milk from its mother and spends most of the summer days running, jumping and sleeping.

A new set of antlers

Every spring, the stag's antlers drop off. You can sometimes find old antlers on the ground. New ones grow very quickly. At first they are covered with a soft skin, like velvet. By the time autumn comes, the skin has dried and the stag rubs it off on trees. Young stags have small antlers. Each year the new ones are larger, until the stags grow old.

The Panda

The Giant Panda looks like a black and white teddy bear wearing sun glasses. It is a large animal, and may weigh over 100 kg. It lives in the mountain forests of China and nowhere else, except in zoos. Although everyone knows what it looks like, the Giant Panda is a rare animal. Only a few hundred are left in the wild.

There is another kind of animal called a panda, the Red Panda. It is nothing like the Giant Panda, and is not even a close relation. It is a cousin of the racoon, and weighs only about five kilograms.

An extra claw

The Giant Panda has six claws on each paw. The extra claw helps it grasp and break off the stems of bamboo.

The Red Panda

This small animal has a tail which is almost as long as its body. It spends most of its time in the trees. It finds a comfortable hollow tree trunk, and sleeps there most of the day. At night it goes hunting for small animals and young birds, but it also eats fruit and leaves.

Bamboo

The Giant Panda will eat almost anything, even meat. But its main food is the leaves and stems of bamboo. In the summer, there is plenty of bamboo where the pandas live. But in the winter the bamboo plants lose their leaves. Giant Pandas may get very hungry before the bamboo starts to grow again in the spring.

The Wild Boar

Boar is the name for a pig. Wild boar and farmyard pigs looked alike a few hundred years ago. Now they look quite different, but it is the farm pig that has changed. There are no wild boar left in Britain or Ireland, and you do not often see one in the forests of Europe, where they still live. They are quite shy, and they lie low in the daytime.

Wild boar are covered with dark, wiry hair. The piglets are brown, with dark stripes.

The boar's tusks

Wild boars look fiercer than farm pigs. They have heavy shoulders, long snouts and they are armed with two tusks. The tusks are short, but very sharp, and the boar uses them in fighting, if it has to.

Feeding

Wild boar snuffle about the forest looking for roots, beech nuts and, best of all, acorns. Sometimes they will raid a farmer's fields for potatoes or sweetcorn.

Wild boar like the dark

Wild boar live in groups. By day, they hide themselves in the forest. When night comes, they go out to look for food.

Families are large. There may be 12 piglets in one litter.

Red Squirrel and Dormouse

In England, the red squirrel has been driven out by the bigger grey squirrel. There are plenty still in Scotland and other countries in Europe. These bushy-tailed animals live in the trees. They can climb a big tree in a few seconds. It is hard to see a dormouse. They are tiny animals, smaller than most other mice, and they spend the day curled up asleep and out of sight. They feed on nuts, and forest fruits. Like squirrels, they eat sitting up, holding their food in their paws.

Squirrels and dormice belong to the family of animals called rodents. They have long front teeth to gnaw and nibble.

A cosy nest

The squirrel makes a nest in a tree out of leaves and twigs. The babies do not stay in the nest for long. They are soon scuttling about the branches and quickly learn to find their own food.

The squirrel is an acrobat

The squirrel leaps so quickly from one branch to another that it almost seems to fly. Its sharp little claws help it cling to the bark.

Nuts about nuts!

Nuts and acorns are the squirrel's favourite food. It also nibbles at buds, young leaves and wild mushrooms.

The sleepy dormouse

When winter comes, the dormouse rolls itself up into a ball in its nest. It sleeps nearly all the time, until spring arrives. This is called hibernation.

The squirrel also hibernates. First, it collects a store of food which it hides in a hollow tree, so it does not have to search for food in the frosty weather. It sleeps a lot, but comes out on mild days, even in midwinter.

The Fox

The fox is about the same size as a dog, and looks rather like a dog too. It has large, sharp ears, a long muzzle or nose, and a lovely long and bushy tail. Although it is an animal of fields and forests, it also lives in towns. Its home is a large hole in the ground, called an earth. In the day, it is often sleeping in its earth. At night it is out looking for food. A town fox will even raid a dustbin when looking for its supper. The fox is famous for its cunning. It can open the door of a rabbit hutch with its teeth. If you have pet rabbits, make sure their hutch is bolted!

Chicken thief

A fox likes nothing more than a fat chicken. When the farmer is asleep, the fox raids the hen house.

Young foxes are called cubs.

After they are born, they stay in the earth close to their mother and feed on her milk. In a few days, they are big enough to play rough games outside the earth. They learn from their parents how to hunt and hide. When they grow up, they move away to make a new earth of their own.

The fox is not faddy

The fox hunts small animals such as mice, voles and rabbits. It eats birds and insects, if it can catch them. It will also eat fruit. In fact, the fox will eat almost anything, including scraps of food it finds in a dustbin.

The Koala

This cuddly animal comes from Australia. It has big, furry ears and a big nose, which give it a comical look. It used to be hunted for its thick, woolly fur, but these days it is protected.

It lives among the eucalyptus forests, and moves around the trees rather slowly.

It is sometimes called a koala bear, and it does look like a small bear. But the koala is not a bear.

Food and drink

The name koala means 'no drink'. Koalas don't drink. They feed mostly on the leaves of eucalyptus trees. They eat more than one kilogram of leaves each day. As the leaves are very juicy, koalas never need to drink water. When the young koala begins to eat leaves, its mother chews them first.

The baby koala

When it is born, the baby koala is smaller than a mouse. It has no hair and it is blind. For the first six months of its life, it never comes out of the pouch in its mother's stomach. Then it is ready to face the world.

When it grows too big for the pouch, it rides on its mother's back.

Asleep in a tree

When the koala goes to sleep, it finds a fork in a tree where a branch meets the trunk. It sits on the fork and digs its claws into the bark. In this position, it sleeps comfortably. It hardly ever falls off.

The Bear

Everyone knows what a bear looks like. It is a big animal, a VERY big animal! But some bears are bigger than others. A giant grizzly bear from North America weighs more than half a tonne. The Malay bear, which lives in the forests of south-east Asia, weighs only 40 kg. The pictures here show the European brown bear, which weighs about 150 kg.

The bear moves about on all fours. Although it is so heavy, it can run quite fast when it needs to. Bears can also stand upright on their short back legs.

Up the tree

Bears may look clumsy, but a young brown bear can climb a tree easily. It grips the tree with its strong claws. Adult bears climb about the rocks just as easily.

Going fishing

Brown bears eat all kinds of food, mostly plants, but also meat if they can find it. They sometimes hunt smaller animals, but they are too slow to be good hunters. Wild honey is a special treat, and they also like fish. A young brown bear can catch a fish with one swipe of its paw.

Young cubs at play

Cubs are born in winter. They live with their mother in a den and don't come out until the weather gets warmer. The she-bear is a very good mother, and sometimes joins in their games.

The Elk

The elk is the largest member of the deer family, even bigger than the red deer (page 6). It lives in cold northern countries. It usually lives near water, or even *in* the water, for elk are good swimmers. They are not so fond of snow, and in the winter they get together in large groups and make 'yards'. These are special places where the elk have trampled the snow flat.

Elk live in North America, as well as northern Europe and Asia. But in North America they are called moose.

The young elk hides from trouble

A male elk is called a bull and a female a cow, so the young elk is called a calf. When a calf is born, its mother hides it in the boggy grass, where wolves and pumas will not easily find it.

Wolf attack

Wolves are the enemies of the elk. They will attack a stray elk which is separated from the herd. The strong antlers and sharp hooves of the elk usually drive off the wolves. But in winter, when the snow is deep and the elk has shed its antlers, the wolves may win.

Finding food

Elk feed on water plants. They are especially fond of water lilies, and eat the stems and roots as well as leaves. In winter, when the water is frozen, food is not so easy to find. The elk may have to live on pine needles.

Only the bull elk has antlers. He also has a kind of beard behind his chin.

The Marten

The marten is a member of the family of weasels. Like all weasels, it is a fierce hunter. It is an animal of the trees. It makes its nest in a hollow, and hunts squirrels among the branches. It leaps from branch to branch with great speed, and can squeeze its body into the narrowest cracks in rocks or tree trunks.

The sable

The sable is a kind of marten, which lives in cold, northern countries. It is famous for its beautiful, dark grey fur.

A fierce hunter

The marten is not often seen in daytime. When the sun goes down, it leaves its hole to begin the night's hunting. It is so quick that it can catch a squirrel in mid-air, but it sometimes hunts on the ground too. It is afraid of nothing, and will kill rabbits and other animals larger than itself.

Baby martens

When the young martens are born, they are very small and helpless. They stay close to their mother in the nest. When they are a few months old, they start to copy their mother as she moves up and down the branches.

The Badger

The badger has short legs, which make it look like a small animal, but it is heavier and stronger than it looks.

It has strong claws, for digging, but it cannot see very well. Instead, it has a very good sense of smell.

Badgers are timid animals, and you may live near them without ever seeing one. They spend the day in their underground home, which is called a sett.

The badger in winter

In the cold winter months, the badger is not so busy. It does not sleep all winter, but it spends many days at a time inside its sett. Because it is not running about the woods all night, it needs less food.

Night-life of the badger

When the sun goes down, it is time for the badger to leave its sett. If the moon is bright, it may stay at home. If not, it sets out to find its food. It is not a great hunter, but it can dig out a mole with its powerful claws. Its favourite food is earthworms. It can gobble up a hundred worms in one night. But the badger is not just a meat-eater. It also eats fruit and berries.

Young badgers

Baby badgers are born when winter is nearly over. By the time the warm weather comes, they are big enough to play outside the sett. The young ones live with their family for about a year. Then they set off to find a new home of their own.

The Owl

The owl is a large bird and is easy to recognise. It has a big, round head and large eyes. Unlike other birds, it has a flat face with eyes in front, and it seems to stare at you. But it is not easy to see an owl, because they are birds of the night. They are hunters, with strong talons, or claws, for seizing small animals like mice. They are more often heard than seen. Enter a wood at dusk, and you may hear an owl hooting, 'Too-whoo, Too-whoo!'

Big eyes

The large eyes of the owl help it to see things a long way off, even when it is nearly dark. It cannot see things close to it so well. As a hunter, it is helped by its good hearing. When flying high above a field, it can hear a fieldmouse moving in the grass far below.

Baby owls are always hungry

Most owls make a nest inside a hollow tree. There they lay their eggs. When the babies are hatched, the mother has to work hard to feed them. Their beaks are always open. She brings them insects, caterpillars, and sometimes a mouse or a shrew.

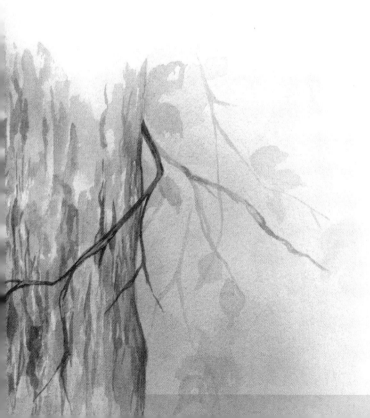

The hunter

The owl flies without making any noise. You cannot hear its wings beating. When it finds a small animal, it swoops down and grabs it in its hooked talons.

There are over 100 different kinds of owls in the world.

CONTENTS

Typeset by TPS Ltd, London

This edition published in 1992 by Regency House
Publishing Limited
The Grange
Grange Yard
London
SE1 3AG

Printed in Italy
ISBN 1 85361 305 3